# BE A SCIENTIST

# LET'S INVESTIGATE ELECTRICITY

## JACQUI BAILEY

CRABTREE
PUBLISHING COMPANY
WWW.CRABTREEBOOKS.COM

# CRABTREE
## PUBLISHING COMPANY
### WWW.CRABTREEBOOKS.COM

**Author:** Jacqui Bailey

**Editorial director:** Kathy Middleton

**Series editor:** Julia Bird

**Editor:** Ellen Rodger

**Illustrator:** Ed Myer

**Packaged by:** Collaborate

**Proofreader:** Petrice Custance

**Production coordinator
  and Prepress technician:** Ken Wright

**Print coordinator:** Katherine Berti

---

**Library and Achives Canada Cataloguing in Publication**

Title: Let's investigate electricity / Jacqui Bailey.
Other titles: Investigating electricity
Names: Bailey, Jacqui, author.
Description: Series statement: Be a scientist | Previously published
   under title: Investigating electricity. | Includes index.
Identifiers: Canadiana (print) 20200353756 |
   Canadiana (ebook) 2020035390X |
   ISBN 9781427127716 (hardcover) |
   ISBN 9781427127778 (softcover) |
   ISBN 9781427127839 (HTML)
Subjects: LCSH: Electricity—Juvenile literature. |
   LCSH: Electricity—Experiments—Juvenile literature.
Classification: LCC QC527.2 .B35 2021 | DDC j537—dc23

**Library of Congress Cataloging-in-Publication Data**

Names: Bailey, Jacqui, author.
Title: Let's investigate electricity / Jacqui Bailey.
Description: New York, NY : Crabtree Publishing Company, 2021. |
   Series: Be a scientist | Includes index.
Identifiers: LCCN 2020045053 (print) | LCCN 2020045054 (ebook) |
   ISBN 9781427127716 (hardcover) |
   ISBN 9781427127778 (paperback) |
   ISBN 9781427127839 (ebook)
Subjects: LCSH: Electricity--Juvenile literature.
Classification: LCC QC527.2 .B284 2021  (print) |
   LCC QC527.2  (ebook) |  DDC 537--dc23
LC record available at https://lccn.loc.gov/2020045053
LC ebook record available at https://lccn.loc.gov/2020045054

---

## Crabtree Publishing Company
www.crabtreebooks.com    1–800–387–7650

**Published in 2021 by Crabtree Publishing Company**

First published in Great Britain in 2019 by Wayland
Copyright © Hodder & Stoughton, 2019

All rights reserved. No part of this publication may be reproduced, stored in a retrieval system or be transmitted in any form or by any means, electronic, mechanical, photocopying, recording, or otherwise, without the prior written permission of the copyright owner.

The text in this book was previously published in the series 'Investigating Science'

Printed in the U.S.A./122020/CG20201014

Every attempt has been made to clear copyright. Should there be any inadvertent omission please apply to the publisher for rectification.

**Published in Canada**
Crabtree Publishing
616 Welland Ave.
St. Catharines, Ontario
L2M 5V6

**Published in the United States**
Crabtree Publishing
347 Fifth Avenue
Suite 1402-145
New York, NY 10016

# CONTENTS

What is electricity? 6

How do we get electricity? 8

Packets of power 10

Right way, wrong way 12

Going around in circles 14

Will it work? 16

On and off 18

| | |
|---|---|
| Test your skill | 20 |
| Flowing along | 22 |
| Brighten up! | 24 |
| Wild electricity | 26 |
| Glossary | 28 |
| Learning more | 30 |
| Index | 32 |

# WHAT IS ELECTRICITY?

**Electricity** is a type of energy. It makes all sorts of things work.

THINK about how most of the machines in your home work.

- You flip or press **switches** to turn room lights on and off.
- You plug in the television to make it work.

How many things in your house are powered by electricity?

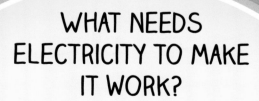

YOU WILL NEED
A sheet of lined paper, a pencil, and a ruler

## WHAT NEEDS ELECTRICITY TO MAKE IT WORK?

1. Draw a straight line down the middle of a piece of paper.

2. In the left-hand column, list all the things in your house that use electricity.

## WHAT IS ELECTRICITY?

> **BECAUSE...**
> We use electricity because it is a handy source of energy. It is easily turned on and off so it can be used only when it is needed.

**3** In the right-hand column, write down what each machine does. Ask an adult if you are not sure.

**4** Group the machines on your list according to what they do. For example, do they give out light or heat, or sounds and pictures? Do they make something move?

# HOW DO WE GET ELECTRICITY?

Most of the electricity we use comes from large power stations. This way of distributing electricity is called the **power grid**.

Power lines and **transformers** can be dangerous. It is important not to play near them. You should also never play with sockets and plugs in your home.

**YOU WILL NEED**
2 pieces of paper
A pencil
Scissors
Glue
An adult to help

**THINK** about how electricity travels through the grid.
• It travels to our homes through thick wires called cables.
• From the cables it goes into wires inside the walls and ceilings in our homes.

What happens when the electricity reaches our homes?

## HOW DOES ELECTRICITY WORK IN OUR HOMES?

## HOW DO WE GET ELECTRICITY?

**1** Ask an adult to show you the following things, but do not touch them:
- An electricity meter
- A wall socket or plug
- A light switch
- A table lamp or radio

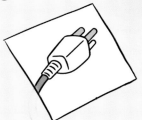

**2** Make a small drawing of each one.

**3** Cut out your drawings. Arrange them in the order in which you think they work to give us electricity. Now, glue them to the second piece of paper in this order.

## " BECAUSE...
We need meters, sockets, switches, and plugs because electricity is very powerful and these devices help to make it safe for us to use. "

# PACKETS OF POWER

Batteries are small devices that supply electrical power.

**YOU WILL NEED**
An alarm clock or stopwatch
1 or 2 friends
Pencils and paper

**THINK** about how we use batteries.
• Batteries fit inside all kinds of small machines, such as watches and cell phones.
• Some machines that use batteries are portable and they can be carried around.

What machines can you think of that use batteries?

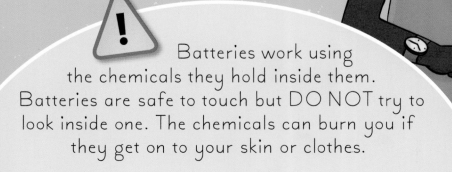

⚠ Batteries work using the chemicals they hold inside them. Batteries are safe to touch but DO NOT try to look inside one. The chemicals can burn you if they get on to your skin or clothes.

# WHICH MACHINES USE BATTERIES?

**PACKETS OF POWER**

**1** Set the alarm clock or stopwatch to ring in five minutes.

**2** In that time, you and your friends can make lists of all the things you can think of that use batteries.

**3** Compare your lists. Who has the longest list? How many different machines do you have altogether?

**4** Look around the house. How many things that use batteries can you find?

## " BECAUSE...

We use batteries because they are easy to carry around and safe to handle. Batteries do not lose any electrical energy until they are used with an electrical machine. "

# RIGHT WAY, WRONG WAY

Batteries have to be put into the machines they power the right way around.

**THINK** about which way around batteries are put into machines.
- Cell phones are designed so that the battery can only fit the right way around.
- Batteries in electric toys have to be put in correctly for the toy to work.

Which way around do batteries need to be?

**YOU WILL NEED**
A flashlight from which you can remove the batteries

## HOW ARE BATTERIES USED?

**1** Turn the flashlight on to make sure it is working.

**2** Remove the flashlight's batteries. Look at them carefully. You should see a plus sign at one end of a battery. This is called the positive terminal. The other end is called the negative terminal. This end should have a minus sign.

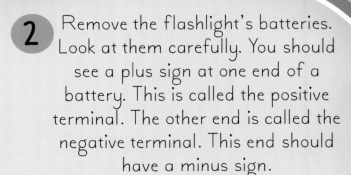

**3** Put the batteries back in the flashlight so that the plus signs are facing each other. Screw the top back onto the flashlight and turn it on. Does it work?

### RIGHT WAY, WRONG WAY

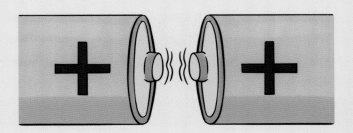

**4** Take the batteries out of the flashlight again. Put them in so that they are facing in the opposite direction to how you found them in step 2. What happens when you turn it on?

**5** Now, put the batteries in as they were when you took them out in step 2. Which way do the batteries' plus and minus signs need to face for the flashlight to work?

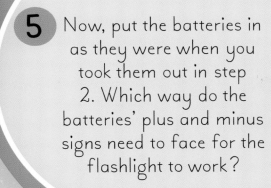

## BECAUSE...

The flashlight worked the third time because the batteries were the right way round. Batteries must be positioned properly in order for them to work. Many battery-operated devices have marks inside. These show the way batteries should be placed.

# GOING AROUND IN CIRCLES

For electricity to work, it has to flow in a loop called a **circuit**. An electric circuit has different parts to it.

**THINK** about how a flashlight works.
- It has a lightbulb, which needs electricity to light it up.
- It has batteries, which provide the electricity.
- It has a metal strip that links the bulb and batteries together so that electricity can flow between them.

This is a circuit. Can you make a circuit?

## YOU WILL NEED
A helpful adult
Scissors
2 lengths of single-strand, plastic-coated wire
A small screwdriver
A bulb holder
A small 3 volt (3V) lightbulb (see page 29 to find out about **voltage**)
A C or AA size battery (1.5V)
tape

## HOW DO CIRCUITS WORK?

1. Ask an adult to strip about 1 inch (2.5 cm) of the coating off both ends of the wires. Twist the ends of the wires to make them neat.

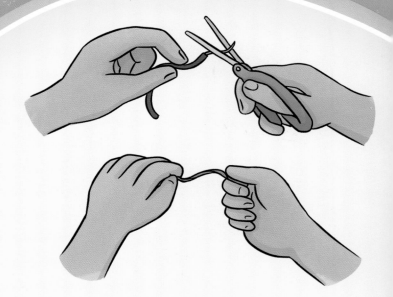

## GOING AROUND IN CIRCLES

> ## BECAUSE...
> The bulb lights up because electricity is flowing from the battery along the wire. It flows through the bulb and back to the battery again in an unbroken circuit.

**3** Carefully screw the bulb into the bulb holder.

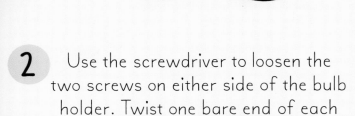

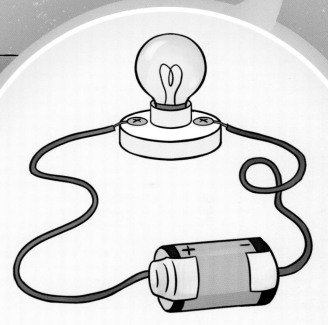

**2** Use the screwdriver to loosen the two screws on either side of the bulb holder. Twist one bare end of each wire around the base of the screws. Tighten the screws again to hold the wires in place.

**4** Tape the other ends of the wires to the positive and negative terminals of the battery. Make sure the bare wires are on the metal contact points of the battery. What happens to the lightbulb?

# WILL IT WORK?

*If any part of a circuit is connected the wrong way, the circuit will not work.*

**YOU WILL NEED**
A pencil and paper
The same circuit parts you used on pages 14–15

**1** Look carefully at the following three circuits.

WHICH CIRCUITS WORK AND WHICH DO NOT?

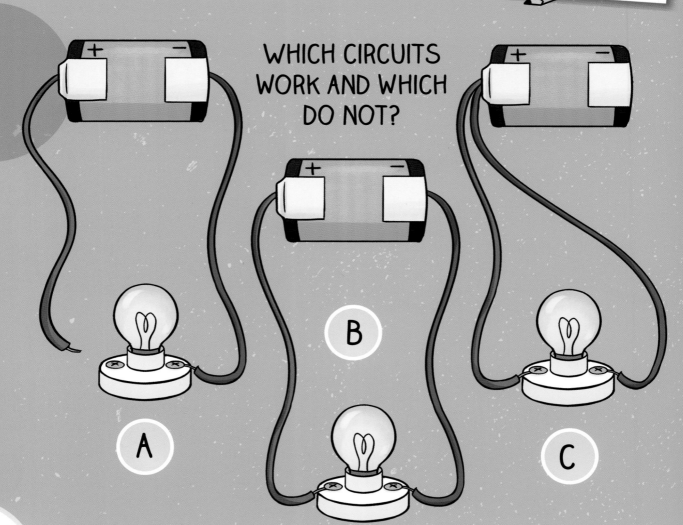

A

B

C

**2** Write down which circuits you think will work and which will not.

WILL IT WORK?

**3** Using the parts from the circuit you made on pages 14-15, make each of the circuits A, B, and C in turn. Were you right or wrong?

" BECAUSE...

Circuit A does not work because the circuit is not complete. There is a break in the circuit between the bulb and the wire. Circuit C does not work because both wires are connected to the same terminal on the battery. Electricity must flow from one end of the battery to the other. Circuit B works because electricity can flow from the battery to the bulb, and back to the other end of the battery. The circuit is then complete.

"

# ON AND OFF

Switches can turn a circuit on and off.

**THINK** about how you turn on electricity in your home.
- You flip a switch on a radio to turn it on or off.
- You press a switch on a computer to turn it on or off.

How do you think switches work?

## YOU WILL NEED

The same circuit parts you used on pages 14–15 and 16–17

2 metal thumbtacks

1 metal paper clip

A piece of balsa wood

An extra length of single-strand, plastic-coated wire

## HOW DO SWITCHES WORK?

**ON AND OFF**

1. Connect your circuit as on pages 14-15, but this time take one of the wires off the battery and wrap it securely around a thumbtack.

2. Push the thumbtack through the paper clip and into the balsa wood.

3. Connect the third piece of wire from the free end of the battery to the other thumbtack.

4. Push this thumbtack into the wood (see left). Make the two thumbtacks the same distance apart as the length of the paper clip.

5. What happens when the paper clip touches just one thumbtack? What happens when the paper clip touches both thumbtacks?

### " BECAUSE...

When the paper clip touches just one thumbtack, the light remains off because there is a gap in the circuit. Electricity cannot flow if there is a gap in a circuit. When the paper clip touches both thumbtacks, the gap is closed and electricity flows around the circuit. The paper clip acts as a switch. Switches open or close a gap in a circuit. "

# TEST YOUR SKILL

Simple circuits can be used in many ways. Test your circuit-building skills by making an electronic game.

## YOU WILL NEED
- Sticky tack
- A wooden board
- A metal coat hanger
- Uncoated fuse wire
- Tape and scissors
- A 4.5V or 9V battery
- A buzzer
- 3 lengths of wire
- A small screwdriver

**THINK** about how each part of the circuit is linked together.

## HOW STEADY IS YOUR HAND?

1. Put a lump of sticky tack onto the middle of the wooden board. Push the hook of the coat hanger into it so that it stands upside down.

2. Loop the fuse wire through the coat hanger and twist the ends into a handle, taping them up to hold it secure.

TEST YOUR SKILL

"**BECAUSE...**
When the loop touches the coat hanger, the buzzer sounds because the loop acts like a switch. When the loop is not touching the coat hanger there is a gap in the circuit and the buzzer does not make a noise."

**3** Connect one wire from the buzzer to the coat hanger, and the second wire from the battery to the buzzer.

**4** Twist the third wire from the battery to the top of the handle of the fuse wire loop. Add some tape to the bottom of the coat hanger as a resting place for the loop. Can you move the loop around the coat hanger without the buzzer sounding?

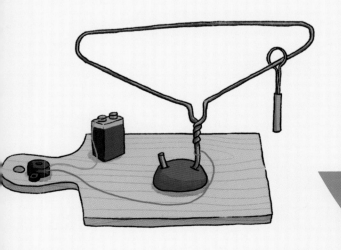

# FLOWING ALONG

Electricity flows through some materials more easily than others. Materials that carry electricity well are called **conductors**. Those that do not are called **insulators**.

## YOU WILL NEED

The same circuit parts you used on pages 18-19, but without the switch
Some test materials (e.g. some wood, a plastic ruler, paper, aluminium foil, a glass, a metal paper clip, an eraser)
A pencil and paper

**THINK** about the materials you use to make circuits.
• Wires carry the flow of electricity around a circuit.
• Wire is made of metal wrapped in plastic.

Which of the test materials are conductors and which are insulators?

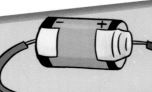

## WHICH MATERIALS MAKE GOOD CONDUCTORS?

1. Set up a circuit like the one on pages 18-19, but this time leave the switch out so that you have two loose ends of wire.

**2** Touch the two ends of wire together to make sure your circuit is working properly.

## FLOWING ALONG

**3** Now touch the two ends to each of your test materials in turn.

**4** Write down each material you use and whether or not the bulb lights up.

> ## BECAUSE...
> The bulb lights up with metal objects because metal is a good conductor. It does not light up when the circuit includes materials such as plastic or wood. These materials are good insulators. Why do you think electric wires are coated in plastic?

| Material | Lights up? |
|---|---|
| eraser | no |
| glass | |
| pebble | |
| foil | |

23

# BRIGHTEN UP!

A circuit can have as many parts, or **components** attached to it as you like.

**THINK** about different types of circuits.

• Christmas tree lights are all part of one circuit. When the circuit is plugged in, all the lights come on at the same time.

What happens when you add parts to a circuit?

**YOU WILL NEED**
4 lengths of plastic-coated wire
A 1.5V battery
3 x 3V-4.5V bulbs and holders
Tape
Scissors
A small screwdriver
A 9V battery

**1** Build a circuit using two lengths of wire, a 1.5V battery, and one bulb, as on pages 14-15.

**2** Add another length of wire and another bulb to your circuit. What happens? Do both bulbs glow?

**BRIGHTEN UP!**

**3** Replace the 1.5V battery with the 9V battery. Does this make a difference?

**4** Add the fourth length of wire and a third bulb. What happens now? Are all the bulbs lit? Are they brighter or dimmer?

"
## BECAUSE...
When you add more bulbs to a low-voltage battery, the lights become dimmer. That's because each bulb gets a smaller share of the electricity. When you use a stronger battery, there is enough electricity for the bulbs to share between them and make them bright. If the battery is too powerful, the bulbs will burn out and stop working.
"

# WILD ELECTRICITY

Electricity does not come only from power stations and batteries. It also exists naturally as **static electricity**.

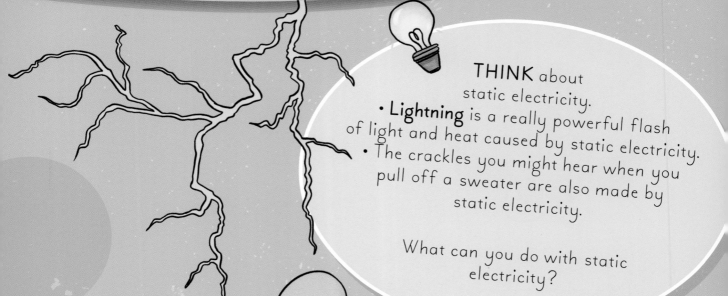

**THINK** about static electricity.
- **Lightning** is a really powerful flash of light and heat caused by static electricity.
- The crackles you might hear when you pull off a sweater are also made by static electricity.

What can you do with static electricity?

## YOU WILL NEED
A blown-up balloon
A wool sweater
Tissue paper cut into small shapes
Test materials (e.g. a wooden spoon, a plastic comb, aluminum foil)
Paper and a pencil

## HOW CAN YOU MAKE STATIC ELECTRICITY?

**WILD ELECTRICITY**

1. Rub the balloon against the wool sweater.

2. Hold the balloon against a wall and let go of it. What happens?

3. Rub the balloon again and hold it near the pieces of tissue paper. What happens to the paper?

4. Try rubbing some other materials on the sweater and hold them next to the pieces of tissue paper. Do any of them affect the paper? Make a record of those that do and those that do not.

## BECAUSE...

"The balloon and some of the other material pull the tissue paper towards it because of static electricity. Rubbing things together makes a static electricity charge build up on the surface. Depending on the material, this electric charge will either pull another material toward it or push it away."

# GLOSSARY

### Batteries
are like tiny portable power plants. They produce electricity by mixing together two chemicals. In some batteries, the chemicals are used up over time and the batteries are then thrown away. There are other batteries that can be recharged and used again.

### Circuits
are the paths that electricity flows around in order to make things work. Electricity must be able to flow in a complete loop from the power source (for example, a battery) around the circuit and back to the power source. If there is a break or gap in the circuit, electricity will not be able to flow.

### Components
are the parts that make up a circuit. Some are devices that need electricity to make them work, such as lightbulbs, buzzers, and motors. Others are the devices that control the flow of electricity, such as switches.

### Conductors
are the materials that let electricity flow easily through them. Metals and water are good conductors.

### Electricity
is a type of energy. We use it to give us light and heat and to power all sorts of machines. Electricity is just one form of energy. Our bodies use food energy to move around and grow. Vehicles such as cars use gasoline for the energy to power their engines.

### Insulators
are materials that slow down or stop the flow of electricity. Plastic and rubber are good insulators.

### Lightning
is a giant spark of electricity that leaps between a cloud and the ground.

## Power Grid
is a network that delivers electricity from generating stations that produce power. The grid includes substations and high-voltage transmission lines. These carry power from distant sources to customers. Generating stations use fuels, such as gas or coal, to make electricity. Hydroelectricity is fueled by moving water.

## Static electricity
is a form of electricity that exists naturally. It does not flow in a circuit, but builds up in materials when they rub together. If enough static electricity builds up, the electricity is forced to flow and jumps as a spark from one material to another.

## Switches
are used to control the flow of electricity in a circuit. A switch turns a circuit off by opening up a gap in the circuit. Switches allow us to save electricity by using it only when we need to.

## Transformers
are devices that change the voltage of an electric current.

## Voltage
is a way of measuring how strongly an electrical flow is pushed around a circuit. The strength of the push is measured in volts. A small battery may have 1.5V (1.5 volts) stamped on it. A stronger battery may have 9V. If more than one battery is used in a circuit, (for example, three 1.5V batteries) the number of volts adds up to a greater voltage (3 x 1.5 = 4.5 volts).

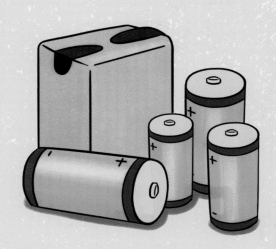

# LEARNING MORE

### BOOKS
Arbuthnot, Gill. *Your Guide to Electricity and Magnetism*. Crabtree Publishing, 2017.

Arnold, Nick. *Horrible Science: Shocking Electricity*. Scholastic, 2018.

### WEBSITES
Learn some cool facts about electricity at:
www.coolkidfacts.com/electricity-facts/

www.dkfindout.com/uk/science/electricity/ has lots of helpful information on batteries, circuits, and even electric animals! It also includes an electricity quiz.

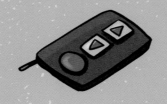

## PLACES TO VISIT
The SPARK Museum of Electrical Invention in Bellingham, Washington, has hands-on learning activities at the museum and and STEM challenges online at www.sparkmuseum.org.

## NOTE TO PARENTS AND TEACHERS:
Every effort has been made by the publisher to ensure that these websites contain no inappropriate or offensive material. However, because of the nature of the Internet, it is impossible to guarantee that the content of these sites will not be altered. We strongly advise that Internet access is supervised by a responsible adult.

# INDEX

batteries 10, 11, 12, 13, 14, 15, 17, 19, 21, 25, 26, 28, 29
buzzers 21, 28

cell phones 10, 12
chemicals 11, 28
circuits 14, 15, 16, 17, 18, 19, 20, 21, 22, 23, 24, 25, 28, 29
components 19, 24, 28
conductors 22, 23, 28

devices 13, 28

electric charge 27
electric flow 14, 15, 17, 18, 19, 22, 28, 29
electric meters 9
electronic games 20
energy 6, 7, 11, 29

flashlights 12, 13, 14
fuels 28

Insulators 22, 23, 29

light and lights 6, 7, 18, 19, 24, 26, 29
lightbulbs 14, 15, 17, 23, 25, 28
lightning 26, 29

machines 6, 7, 10, 11, 12, 29
materials 22, 23, 27, 28, 29
metals 14, 22, 23, 28

negative terminals 12, 15

plastic 22, 23, 29
plugs 6, 8, 9
positive terminals 12, 15

power grid 8, 9, 28, 29
power stations 8, 26, 28

radios 9, 18

sockets 8, 9

static electricity 26, 27, 29
switches 9, 13, 18, 19, 21, 22, 28, 29

voltage 14, 25, 29

wires 8, 9, 14, 15, 17, 19, 20, 22, 23, 25, 28, 29

32